"The Beneficent Agencies of the Useful Arts."

# MR. DEMING'S SPEECH

FOR THE

# Useful Arts.

# A SPEECH

# FOR THE USEFUL ARTS.

DELIVERED AT

NEW HAVEN, JANUARY 2d, 1856,

BEFORE THE

Agricultural Society of the State of Connecticut.

---

BY

HENRY CHAMPION DEMING.

---

HARTFORD:
PRESS OF CASE, TIFFANY AND COMPANY.
1856.

"WHATEVER ELSE MAY TEND TO ENRICH AND BEAUTIFY SOCIETY, THAT WHICH FEEDS AND CLOTHES COMFORTABLY THE GREAT MASS OF MANKIND SHOULD ALWAYS BE REGARDED AS THE FOUNDATION OF NATIONAL PROSPERITY."—WEBSTER.

"HE THAT HATH A TRADE HATH AN ESTATE. HE THAT HATH A CALLING HATH AN OFFICE OF HONOR AND PROFIT."—FRANKLIN.

HARTFORD, January 12th, 1856.

Hon. HENRY C. DEMING:

SIR:—At the meeting of the Connecticut State Agricultural Society, January 2d, 1856, it was

*Resolved*, "That the thanks of the Society be extended to the Hon. Henry C. Deming, for his able and eloquent address, and that he be requested to furnish a copy of the same for publication with the Transactions of the Society."

By order of the Executive Committee, I have the honor to transmit the above resolution, and to request your manuscript for publication.

Very respectfully,

Yours truly,

HENRY A. DYER, *Cor. Sec'y.*

---

HARTFORD, January 15, 1856.

HENRY A. DYER, Esq.:

Dear Sir:—I have received your letter, containing the resolution of the Agricultural Society, and asking a copy of my address, for publication. A manuscript had been written, containing a small portion of what I had intended to say on the State Fair Grounds, if that intention had not been defeated by the rain. When I was unexpectedly called upon to go to New Haven, and deliver in a January ice-storm, an address which had been prepared to be spoken "under an October sun," and in the actual presence of your Exhibition, I was obliged to abandon much of my previous preparation, and the sheets which I now send you, are in part a report of the New Haven speech, written out since the second of January, and in part such fragments of the manuscript as I found available for that occasion.

I am, very respectfully,

HENRY C. DEMING.

# SPEECH.

---

Mr. President:

The Second Annual Fair of your Society, as proof and profert, of the agricultural and industrial wealth and resources of the State, might well command our pride and gratitude. While the display of Durhams and Devons, was highly creditable to the breeders of fancy stock, there was an array of native breeds and working oxen, representing the most real and substantial uses of this animal, that was, by universal consent, not only entirely unsurpassed in all previous exhibitions of a similar character, but so large and so rich, as to safely challenge and defy the future.

As the commerce of a State is measured by the tonnage of its ships, so the number and condition of its Working Oxen, may be regarded as no unfair exponent, of its agricultural strength and capacities.

We had horses, too, of every variety, natives and thorough-breds, for show and for use; coursers and roadsters, which in beauty, strength, and speed, compared favorably with the collections drawn together, by the larger premiums, and broader competition of National Fairs, devoted to this animal exclusively. While we had steeds that would enchant an Arab, it was my good fortune to notice one pig, that might have converted a Turk; but he stood solitary and alone in his swinish loveliness ; the only beauty of his family that graced the occasion.

If a more delicious and fragrant offering, can be laid on the altars of Bacchus and Pomona, than was to be found in your horticultural tent, I trust that I, and all lovers of good fruit may be there to see it.

The Mechanical and Manufacturing Department of the Fair, though commendable, so far as it went, and presenting many beautiful fabrics, and much exquisite workmanship, yet as a whole, was an inadequate exhibit of our wealth in these departments, and immeasurably beneath our capacity. For if there is one State that might throw down the gauntlet to the world, in such an exhibition, that State is Connecticut. If there is one State where mechanics and manufactures are not confined to the water-courses, and the sides of the railroads, but like the atmosphere, pervade and permeate every part, that State is Connecticut. If there is one State, distinguished no less by those great establishments that throw off their products by the hundred thousand dozen, than by those little rills and streams of industry, that sparkle and ripple on every homestead, that State is our own. If but a tithe of these could have been diverted from their channels, and turned into your enclosure, there would have been such a numberless and nameless variety of contrivances and notions, such a museum of every conceivable and inconceivable patent, and labor-saving machine, such a gathering together of the multitudinous progeny of Connecticut cuteness, and cunning, and skill, as the sun in his course, never looked down upon.

You can hardly expect from me, sir, on this occasion, or on any occasion, any instruction in practical husbandry, or in agricultural art and science. Neither my studies nor pursuits, have adapted me to any such undertaking. The pleasures of the farmer's life! have they not been said and sung from the days of Theocritus and Virgil, to the last orator of the day? I should despair attempting to decorate a theme, already so amply illustrated, or to increase the interest we all feel, and the respect we all entertain, for the healthy, honest, and honorable vocation of those, who, masters of their own acres, as well as of their own wants and passions, are the truest exemplars of republican frugality, independence, and manliness. Although, sir, under the auspices of your Society, and although agricultural productions and implements constitute a large part, still your exhibitions are intended to display all the industrial arts and resources of the State, and it is this aspect of them, that naturally suggests the general theme, to which I have ventured, with many misgivings, to call your attention, for a few moments, to wit:

THE BENEFICENT AGENCIES OF THE USEFUL ARTS.

I am aware that Agriculture is not included in those classifications of the USEFUL ARTS, that are merely scientific; but as I know no good reason for the omission, and am moreover aiming at no severe distinction, but seeking a convenient term, to embrace most of the forms of industry, represented in your Exhibition, I shall use the term, as if Agriculture was confessedly within its purview.

Among the instrumentalities which affect the condition of man, the precedence is quite uniformly, and rightfully given to those, which address themselves to his spiritual nature, to Religion, Education, Law, and the Fine Arts. But if these agencies were the first, in order of time, as they are in rank, to which he is subjected, and after they had done for him their utmost, he should be bereft of others, equally indispensable to his welfare, he would find himself the most miserable and pitiable specimen, of the mamalian family.*

He might be good, wise, upright, "noble in reason, infinite in faculties, in form and moving express and admirable, in action like an angel, in apprehension like a god," but he would be a naked, thin-skinned, hungry, thirsty, short-winded, shame-faced biped, without hide, fur or feathers. In such a condition, the USEFUL ARTS receive the paragon of animals, from the hands of his spiritual guardians. They feed, clothe, shelter, cleanse, adorn him. In comparison with other creatures, they find him weak, and endow him with a strength

---

* Bishop Potter, in his valuable little treatise "on the Principles of Science applied to the Domestic and Mechanical Arts," says, in substance, "that the difficulty with which man maintains even life, when deprived of some of his customary tools and weapons, and left without the assistance of his fellows, is strikingly illustrated by the adventures of Ross Cox, who, while traveling with a company of traders, in the North-West Territory, was one day, when he had fallen asleep, accidentally left by his companions, and not found again till several days after. It happened, that owing to the extreme heat of the weather, he had divested himself of his weapons, and of nearly all his clothing, which was carried off by the party. He was left, therefore, like other animals, to his natural resources; and the picture which he gives, of the extremities to which he was quickly reduced, by hunger, by the torturing stings of insects, the terror of wild beasts, and the impossibility of tracing his companions, is equally affecting and instructive. It is quite evident, that, if he had not been most opportunely found, he must soon have perished." In a note to the above passage, he adds, "It may be possibly objected, that this example of man's natural imbecility, is taken from an individual who has once lived in civilized society, and had thus been rendered effeminate; and that we ought rather, to adduce the case of those who have lived, like animals, only in a state of nature. One or two such individuals have been discovered, within the last century, living alone, in forests; as for example, Peter, the Wild Boy, in Germany, and the savage of Arvignon, in France. They were found, however, in the lowest physical and intellectual condition, subsisting on the bark of trees, unable to distinguish, by touch, between a carved and painted surface, and having hardly a trace of humanity. Such cases afford the most conclusive proof, that, as *men*, we owe most of our power and dignity to culture.

superior to all; defenseless, and equip him with arms that vanquish all; slow, and give him wings that outstrip the eagle; in short, they encircle his perishable, with comforts and luxuries, worthy of his imperishable nature.

Though the Useful Arts find Man thus destitute personally, he is no beggar, but the undoubted and rightful heir of a most splendid inheritance,—useless and unavailable, it is true, in its normal condition, but under proper culture and management, an inexhaustible mine of plenty and wealth. It consists of the rough matter which composes the solid earth; of the soil and water which cover it; of the birds of the air, the beasts of the field, and the fishes of the sea. Of this inheritance, the Useful Arts become the most serviceable and trustworthy of stewards. They convert the solid earth into innumerable objects of convenience and value. They open communications, secure the harvests, collect the flocks, cultivate the soil, improve the fisheries; in short, they render the world, over which dominion was given to man, a comfortable, convenient, and elegant abode.

God creates matter, but the Useful Arts create its utility, or, in the language of Political Economy, they are producers, and production is the sole, the only fountain head, of that enviable stream, the Wealth of Nations. Commerce, to be sure, is an important agent, in diverting the current, and in changing the relative position of wealth, but it adds not one drop to the golden stream, for which countless myriads thirst. Production is its only origin, and every flight of human credulity, every device of human ingenuity, to discover some other source of this Pactolus, has signally failed. The Golden Fleece, the dreams of the Alchemist, the visions of El Dorado, South-Sea Bubbles, tulip-manias, multicaulis-manias, California fevers, stock-jobbing, and up-town lots, are the weighty authorities and confirmation strong, which successive centuries have brought to the truth, that production is the only real source, of the aggregate wealth of Nations. Many falser images have been used than that which declares, that "Gold, in its last analysis, is the sweat of the poor and the blood of the brave."

It is a liberal estimate, which assigns one-fifth of the human family, in civilized countries, to the non-producing class; the USEFUL ARTS provide for the remaining four-fifths, and thus convert into props, and pillars, and bulwarks, what would otherwise be, intolerable drags and burthens and nuisances, in a State. They give employment, not servile and degrading, but honorable and remunerative employment, to a vast majority of the human family. This consideration alone, if it was all that could be urged, would place them foremost, among the agencies which contribute to the welfare of the race.

But still higher commendation belongs to them. They are the grand instruments, by which LABOR acts upon the world, and thus the paramount obligations justly due to LABOR, become justly due to the USEFUL ARTS. "In the sweat of thy brow thou shalt eat bread," is a curse which carries a blessing with it. Like Mercy, labor is twice blessed,—

"It blesseth him that gives and him that takes."

That toil to which we are condemned, as the tenure of existence here below, is the training, which invests both body and soul, with the insignia of true and genuine manhood. Effort is the only school for the muscles of the frame, and the muscles of the intellect. Where but in that rocky mine which LABOR delves, can be found those priceless gems, will, efficiency, courage, pluck, perseverence, patience, self-confidence, self-reliance, contempt for difficulties? These are the sheet-anchors of the heroic character,—this is the stuff of which martyrs and heroes are made,—these fashion those souls, that are adamant in a just cause. Goethe gracefully compares the effect, of a strong necessity, imposed upon a mind, habitually untasked, to an oak planted in a China vase; when the branches expand and the roots strike out, the vessel flies to pieces.

Invaluable as is this disciplinary function of LABOR, it is but a pebble picked up on the shore—a drop in the boundless ocean of her beneficence. LABOR is a universal solvent, a philosopher's stone, with transmuting powers, magical and gorgeous, beyond the dying alchemist's dream. Entering into all the dead, sluggish, inert matter of the earth, she imparts to all the life-like properties of Utility and Value. There is nothing in the caverns of this round globe, in the depths of the sea, I had almost said the realm of Air, which LABOR transforms not into a necessity or a luxury. No sweep of ocean, no forbidding desert, no fastnesses of forest or of wilderness, can hide a product, useful to man, from the omnipresent eye of this great benefactor. She catches from the passing breeze, the waste white down of the cotton shrub, and lo! bleaching cloth lies in the place of idle litter, and the nakedness of man is covered.* She stumbles upon a worthless mass of vitrified sand, and behold! window-panes for every man's dwelling, cheap drinking cups for every man's table, the mirror, the Portland vase, the prism, the telescope, the microscope, the Crystal Palace. It stretches its hand over the waste places of the earth, and "instead of the thorn, comes up the fig-tree, and instead of the brier, the myrtle-tree." Iron, in its fingers, is as flexible as clay in the potter's, while language struggles in vain to depict, the infinite variety of texture, and utility, which it imparts, to the fleece of an animal, the gum of a tree, and the entrails of a

*Carlyle.

worm. There are no such words as "useless," "worthless," in her vocabulary. Refuse and rubbish are no longer such, when touched by her wand. The dead animal, which was formerly banished to the wilderness as a nuisance, she now transmografies into something useful or ornamental; she even brings life out of death, vitalizing exhausted soils, by the moldering relics of mortality, which she digs from the Blenheims, the Austerlitzes and Waterloos of the world.

The blessings which mankind owes to PRODUCTIVE LABOR, can be vividly realized, by imagining the state of things, if it should be annihilated. Suppose, then, by some all-pervading distemper, or by some fiat of divine displeasure, the arm of universal LABOR was paralyzed. It would break the main-spring which sets the whole machinery of existence in motion. It would cut off the supply of life at the fountain. The wheel of business, losing its only momentum, would soon cease to revolve. Grass would grow in our most crowded thoroughfares. Those great marts, where traffic now chaffers in its thousand tongues, where cheerful art rings its innumerable sounds, and busy and hurrying myriads proclaim the bright and joyful reign of LABOR, would become noiseless, and blighted, and petrified, like some vast city of the dead. Not the clink of a hammer, nor the rattle of a shuttle, nor the whiff of a steam-engine, nor the roll of a wheel, would break the sepulchral stillness of an idle world. The axe, the file, and the saw, would lie silent, where they had dropped from the hands of the yawning artizan; the plough would rust where it had stopped in the furrow. All the products of the now idle weaver, would soon drop piecemeal from the shelves of the merchant, and tattered rags, hanging on a universe of sluggards, would pre-announce man's speedy return, to his original Nakedness. Crops would decay in the field. The ungathered fruits would rot upon the ground; the granary would soon surrender its last kernel. Starvation would follow Nakedness. Ships, sailor-less, would toss upon the seas, the forests would be burnt for fuel, the mine would no longer send to our wharves the grateful coal, and Frost—a third fury—would follow in the footsteps, of Nakedness and Famine. The palaces of the great, the habitation of every family, would be burnt for fire, whole cities would be consumed, and naked and starving man would soon be houseless, shelterless, and gathering round the dying embers of their dwelling, would rake together the feeble sparks, with skeleton fingers. Religion, Education, Law, the Church, the Altar, and the Capitol, would all be whelmed and wrecked in a world-wide maelstrom, of wretchedness and despair.

If it is true that the USEFUL ARTS are thus of indispensable necessity to mankind, it would be reasonable to suppose, that they had always been, the special objects of supervision and care, on the part of the science and learning of the world. So unwarranted, however, is this supposition, that for nearly two thousand years, from the days of Aristotle, at least, to the days of Bacon, there was not only no cordial alliance, between the man of thought and the man of work, between PHILOSOPHY and the USEFUL ARTS, but mutual alienation and distrust. PHILOSOPHY, disdaining to minister to the vulgar wants, or vulgar resources of the race, was absorbed in abstractions, barren and useless to man, as those God-forsaken wastes, which cover the accursed cities of the plain. Her super-refined and exquisite spirit, shrunk instinctively, from the vulgar useful, and the dirty practical. She was ashamed of her accidental improvements, of her involuntary inventions. She disowned and repudiated any chance offspring, which contributed, at all, to the relief of man's estate. She rebuked and denounced one disciple, for having ventured to claim for her, the discovery of the principle of the arch, and the introduction of the use of metals, and ostracized another, for prostituting Geometry to the invention of Machinery, and the most practical of her school, the inventor of the pulley and the endless screw, regarded these contrivances, as scientific trifles, beneath the dignity, and unworthy the severer studies of PHILOSOPHY. "What," said she, in sovereign scorn, "has PHILOSOPHY to do with houses, roofs, iron, copper, ploughs, looms and machinery. PHILOSOPHY raises man above these sublunary wants and expedients, these groanings, sufferings, and complainings, of his mere animal nature. While forming the soul, etherealizing the mind, and lifting both above the visible, the tangible, into the sublimated region of pure thought, and immutable essences, how can she bend her lofty pinion, to brood over mere bodily wailings and wants?" The true philosopher ought not to know, whether he is sheltered, fed, and clothed, cold, hungry, sick, and destitute; Oh! no, he should feel nothing, but the rapture of contemplation, see nothing, but the beauty of virtue, hear nothing, but the music of the spheres! That "there is no reasoning with hunger" or "arguing with pain," are modern fallacies. According to ancient PHILOSOPHY the cravings of a famished stomach, and the wailings of a gouty toe, were both to be cured, by simply convincing their victims, "that pain was no evil." She was eternally in search of the "wise man," and thought that hunger, thirst, pain, the loss of friends, and all the other ills that assail this worthless and contemptible soul-case, could be baffled by some superlative, transcendental, ultra-mundane state of the soul itself. What a perversion here of the name, the ends, and the uses of divine PHILOSOPHY! What a caricature on the benevolent attributes and objects of Wisdom! Advanced as the ancients were, in some sciences, up to the modern standard, and in others possessed of principles of inestimable value, the good of mankind, the physical comfort of the race, the increase of human power, the subjugation of nature's laws, material progress and amelioration, entered not into the dreams of those, who in the Academy and Portico, wasted on abstractions the divine energies of Thought, or wrapped in esoteric exclusiveness, those treasures of Wisdom and Science, God-appointed, like the sunlight and rain, to a beneficence, impartial, undiscriminating, and universal.

When Christianity triumphed over Paganism, these absurdities of the old PHILOSOPHY, were engrafted upon the new religion, and the reign of barrenness and desolation, was prolonged for several centuries, by the Schoolmen squandering on dialectic subtleties, an amount of intellect, hardly inferior to that, which the Platonists had lavished, on these phantasmagoria of brain. While the PHILOSOPHY of the world, was thus on the cloud-capped mountains, of visionary speculation, isolated, insulated, as it were, from the wants and business of daily life, the USEFUL ARTS were toiling on, in the hot and dusty plain, in hand-to-hand conflicts with man's necessities and wants, struggling blindly and recklessly under the heavy load, trying one random experiment after another, without knowledge or principle to direct them, abandoned by their guides, having received no counsel or aid from Science, from the age of Archimedes, to the age of Galileo. Uncared for by Astronomy or Mathematics, during the long night of the Middle Ages, the mariner crept along the threatening lee shore, from headland to headland; the husbandman, in utter darkness as to the constituents of soil, employed no more perfect means of culture, than Jacob might have used, on Laban's farm, and the Mechanic, desponding of further assistance from the Natural Sciences, sank down, in a sort of dogged despair, with such tools and implements, as had descended to him, from the first apprenticeship of the Arts. William the Conqueror led his army, from Normandy to England, with no better means of transportation, than Ulysses had used, in leading his hardy islanders, from Ithica to Troy, and the same kind of weapons, smote our Anglo-Saxon forefathers, with which Joshua smote the Canaanites.

As late as the middle of the fourteenth century, our own ancestors, in "Merry England," were no better skilled in the working of metals, and the weaving of fabrics, than were the inhabitants of forgotten Tyre, and patriarchal Damascus; and in houses, house-hold equipments, and what are now regarded, as the mere necessities of life, they were hardly upon a par with Robinson Crusoe, in his desolate island. Even down to the time of Charles the Second, Dublin was further from London, than Sebastopol is now, and Boston merchandize was longer in reaching Hartford, than it now is in reaching Iowa. In the reign of William and Mary, and of Louis XIV., a French fleet which swept the Channel, is represented, as composed in great part, of galleys resembling rather those ships with which Alcibiades and Lysander disputed the sovereignty of the Ægean, than those which contended at the Nile and at Trafalgar.

Yes, to this ignorance on the part of PHILOSOPHY, of its true mission, the USEFUL ARTS owe a servile bondage, of twice ten centuries, to the blind guides of tradition, and the narrow teachings of personal experience. Under such tutelage, they limped and staggered along the path of improvement. But while thus bereft and at the same time performing unassisted, all the drudgery of society, before the eclipse that darkened and veiled the sun of science had entirely passed away, as an antepast of the blessings they would dispense, when he burst upon them in full-orbed, meridian splendor, they were enabled to give the Telescope to Astronomy, the Mariner's Compass to Navigation, Gunpowder to War, and to Letters, that second savior of the human family, the glorious art of Printing.

It was more than twenty years after the commencement of the seventeenth century, before the Utilitarian Era may be said to have dawned. When Lord Bacon gave to the world a new method of interrogating nature, and declared that the "*end of all science is to enrich human life with useful inventions and arts*," he annulled the long divorce, between PHILOSOPHY and LABOR, harmonized Thought and Work, and thus two seraphs, which since the fall, in separate spheres, had burned and sung, now blended in a common beam, and swelled a common harmony.

What countless blessings have sprung from this auspicious union! When the voice of Spring passes over the frozen North, the silent and ice-locked water-courses, run sparkling and babbling to the sea; the stripped and mournful trees, throw all their gaudy ensigns, and pennants, and fluttering streamers to the fragrant breeze, the air is musical with the liveliest notes of birds, and the living green of the earth, is almost peopled with those fairy folk,—the flowers: in a word, Life springs from Death, as if some divine voice had broke the fetters of the tomb. A change as wonderful, as radical, as universal, as beneficent, consecrated the union between the USEFUL ARTS and the NEW PHILOSOPHY. The night was passed, and the morning broke brightly; the years of famine were ended, and the years of plenty commenced. The thought of the world was turned from theories, dry as the dust of Pharaoh, and from problems, visionary as those, which sought to extract sunbeams from cucumbers, and wholesome food from its digested elements, to theories which were fruit producing, and to problems, which were practicable. For the first time, those four mighty powers, the will, intelligence, skill and strength of the race, united, harmonized, put their shoulders to the wheel of progress and amelioration. Invention followed invention, discovery discovery. Nature began to surrender her secrets and powers. Man, for the first time, began to possess the Earth; for him, the ground brought forth abundantly; for him, Old Ocean smiled serene; for him, wafted on the willing Air, came the "spicy gales of Araby the blest;" to him bowed the winds and the waves, and Steam, Electricity, Galvanism, assumed the yoke of Man, their master, as humbly and willingly, as the ox and the horse.

In the benefits dispensed, by this harmonious co-operation of Science and Labor, no class of mankind *has participated more largely than the laborers themselves.* It has united intellect and work, and wiped out the narrow and conventional distinction between the thinker and the worker, for the worker is now often thinking, and the thinker working, and thus thought has been made healthy by labor, and labor happy and honorable by thought.* In the place of his own limited experience, it has given him, the accumulated experience of all generations of men, reduced to principles, and systematized to laws. It has warned him away from aims that were visionary or impossible, like the pursuit of the perpetual motion, and guided him into paths that would certainly lead to new and valuable adaptation of Nature's laws. It has economized all the forms and processes of labor. It has lifted the intolerable load, which formerly ground him to the dust, precluding all intellectual and moral culture, and placed it upon the untiring muscles and sinews of inanimate machinery. It has sharpened his intellect, enlarged his per-

* Mr. Ruskins' "Stones of Venice," suggests this idea.

ceptions, elevated his position, increased his comforts and enjoyments, improved his health and lengthened his days.

The beneficent influences of this union, *upon the general welfare of mankind*, is of itself a theme which a folio could not exhaust.

We are told of a Geological Era, when the globe was only suited to fishes, saurians, and cold-blooded animals, and that great elemental and organic changes were required, to adapt it to be the dwelling-place of man. Since the harmonious co-operation of Work and Thought, the USEFUL ARTS have completed, what this Geologic revolution commenced, and have finished up the world as a habitation, not only for man the animal, but for man the moral, intellectual, social, industrious, aspiring being. Its heat, frost, lightning, water, darkness, and exhalations, they have rendered innoxious, by equipping him with armor and weapons, that securely defy them all. They have broken through, or leveled down its mountainous upheavals, that Man might circulate broad and free. Its rivers, friths, arms of the sea, and the sea itself, they have bridged, with subaqueous telegraphs and steamships, that to all nations, might be dispensed those blessings, which formerly hovered over a few enviable Edens, and that the interchange of business, and of all friendly offices, between all earth's families, might be as frequent and as facile, as if they were but one community.

Its space and distance, which formerly rendered the different nations, as much strangers to each other, as they are to the inhabitants of Saturn, the USEFUL ARTS have broken down and annihilated. Within one short month, the artisans and scholars of all lands, can now be gathered together in some central Crystal Palace, and in half a month, the armies and the armories of Western Europe be thrown upon the Euxine. Many of its diseases, which once, swept him off in masses, they have extinguished, and have rendered safe for himself, broad expanses of his domain, where before, pestilence and miasmas could only be acclimated. The Montgolfier, the Safety Lamp, and the Diving Bell, have vindicated his right to possess and occupy the Air, the Caverns, and the Sea. In short, Life, Health, Happiness, Comfort, Manufactures, Commerce, Art, Letters, all the great interests of society, have been improved, promoted, perfected, by the USEFUL ARTS, since this new evangel; and I must dismiss this part of my theme, with this imperfect generalization, because this part of it is endless.

Glance now, for a moment, at the *social and political influences* of the USEFUL ARTS, under the new dispensation.

Race, Climate, Soil and Government, were previously recognized, as the grandest instrumentalities, in changing the seats of civilization and empire. But in the august procession of causes, that lead to such magnificent results, they must now yield the precedence to the USEFUL ARTS. The invention of Whitney, adjusted the social position and relations of our Southern brethren, more decisively, than their cotton-perfecting soil and climate. In her Titanic struggle with France, during the first half of the present century, England was as much indebted to Watt, as to Wellington, and to Arkwright, as to Nelson. In peopling and civilizing the wilderness, the institutions of Government, must hide their diminished heads, in the presence of the institutions of Steam. What government, laws, or system, or code of laws, could have done so much for our Mississippi Valley, as the practical development of the great idea which germinated in the brain of JOHN FITCH, of Windsor. It would almost seem, that they were the counter-parts of each other; that the Mississippi Valley was unfinished until the Steamboat was added to it, and that the Steamboat was incomplete until it was stemming the majestic current of the Father of Waters. What woes population in old Connecticut, from the hill-tops to the valleys, like the whistle of the Locomotive? What stretches out the city into the country, and shoves the country on to the city, like the Railroad? What is blending nations together, and blotting out the confusion of tongues, like the Steamship? To what are we indebted for the social unity of the Republic? Primarily, doubtless, to common historic antecedents, to a common language, common institutions and interests. But all these connecting ties have existed, at other periods, and under other conditions, and yet—

> "Lands intersected by a narrow frith,
> Abhor each other, and Mountains interposed,
> Make enemies of nations, which else
> Like kindred drops, had melted into one."

We owe our exemption, from that distrust and alienation, which have hitherto overtaken homogeneous communities, scattered over wide belts of the earth's surface, to the beneficent agencies of the Useful Arts. For the first time in the world's history, friths and mountains, and rivers, and space, are no barriers, to the free and rapid interchange, of thought and feeling. The Railroad, the Telegraph, and the Double Cylinder Press, have so perfected intercommunication, "that as a matter of mere routine," (as it has been strongly and truthfully put,) "the intelligent community from Portland to New Orleans, from Savannah to Chicago, read the same public information, in the same version of it, and the reflecting minds of the whole country are brooding upon the state of affairs in every part, and maturing their judgment concerning it, at the same moment; and when the high debate opens, when the powerful writer or the great orator, utters the developed sentiment of one region, and confirming echo or bold challenge answers from another, proposition and reply, assertion and replication, are alike submitted to the simultaneous judgment of the whole community. It is then no mere fiction or hyperbole, to speak of a nation of twenty millions, as sitting in perpetual council, subjected ever to the same kind of impression, and permeated with the same currents of influence, as swayed the fierce democracy of Athens, driven within the scarlet cords of the Lexiarchs, or govern the sober democracy of New England in town meeting assembled." *

Still higher triumphs, I confidently believe, await the Magnetic Telegraph. In spite of recent experimental failures, it is yet destined to cross the ocean, and to interlace and amalgamate continents and hemispheres, as firmly and closely as it does this sisterhood of States. As many cities will claim the invention of the Sub-marine cable, as claim the birth of Homer. In the remote ages of the future, as far removed from us, as we are from the Ptolemies, it may be the only event, in our day, of sufficient splendor, to render visible to their eyes, that little point in the infinite vault of Time, known as the nineteenth century. I shall be pardoned, I hope, in this connection, if I place thus early upon your records, the claim of my own municipality to this honor. In the winter of 1842–3, a citizen of Hartford—the same who, without loan or discount from the Banks, carries on his single shoulders the tremendous load of our South-Meadow Improvements, but at the time of which I speak, so poor, that he could scarcely call his mathematical instruments, or even his watch his own—he, Sir, laid down on the bottom of the East river, near the line of the Fulton ferry, a Sub-marine

* Oration on "Public Life," by William M. Evarts.

cable, and higher up at Hell-gate another, which differed only from the one, recently lost between Nova Scotia and Newfoundland, in this respect, that in the latter, gutta-percha was used as an insulator, whereas in the former (gutta-percha being then unknown,) a combination of asphaltum and bees-wax was used as an insulator. He actually had in working order, an Electric-Telegraph between Coney Island and the Merchant's Exchange, in which this Sub-marine cable was a *part* of the communication. I, therefore, file my caveat here, in behalf of Col. Samuel Colt, of the City of Hartford, for the invention of the Sub-marine cable.

I have frequently attempted to divine, what would be the emotions of an intelligent mind, if all the surprising revelations of modern art, burst at once upon it, instead of overcoming its special wonder, by their single approach, and slow and gradual development. Suppose the stately form of the Apostolic Hooker,* could rise from yonder grave-yard, and gaze once more on this beautiful valley, how would his mind, richly furnished with all the learning of his time and ingenious and God-trusting, beyond his fellows, regard the changes which two centuries have wrought? Let us meet him, after the first glow of gratitude and admiration, have faded from the manly beauty of his countenance, at finding cities and villages, in the place of the overarching forest, and cultivated farms, instead of an unbroken wilderness. All this, astonishing and wonderful, as indications of progress, were within the range of his anticipations and hopes. Cities as large,

---

*In this speech as originally prepared, I had intended to use "Hooker," as the representative of the past. When it was delivered in New Haven I used the name of "Davenport" as more appropriate to the locality, and varied the sequel accordingly. The text now follows the original manuscript.

villages as flourishing, he had seen before. Let us meet him, when his inquiring mind, is attempting to fathom some modern enigmas, which shock and startle, all the deductions of his science, and all the suggestions of his past experience.

He might expect to find, that even the rich alluvial of the valley, worn out by two hundred years of tillage, had driven his murmuring descendants, into another Exodus, from their ancestral seats. But he learns, that by a process, called the rotation of crops, and by adapting newly discovered fertilizing agents, to the chemical affinities of the various soils, that great mother bosom, the earth, is as unexhausted and prolific, as when he first looked upon its charms, in the sacred solitude of nature. We will say nothing of his surprise, at finding Brindle and Spot, which dragged his wagons from Cambridge, superseded by cattle of blood and lineage, the lordly Devon, and the high-bred Durham, or at finding the stately Saxon and Merinos, in the place of the coarse-wooled dwarfs, or the bow-legged otters, nor of his glance of mingled astonishment and contempt, at the gallinaceous Brobdinags which have invaded the dunghill.

With what unfathomable wonder he surveys those glittering points on chimney-top and steeple, which guide the lightning "innocuously from heaven to earth;" these attenuated wires that flash along the thought of widely separated communities, and searches in vain, for this invisible fluid that "lights up the night with the splendor of day?"

As he gazes round the horizon, he was prepared to find the cattle-path, which led him through the wilderness, grown into the smooth and convenient highway, but he sees an immense causeway, stretching off beyond the reach of his vision on an unbroken level, and on it, (drawn by an agent, whose sole function in his time, was to lift the lid of his dinner-pot, or sing in the nozzle of his tea-kettle) a conveyance, which,

"Singing through the forest,
Rattling over ridges,
Shooting under arches,
Rumbling over bridges,
Whizzing through the mountains,
Buzzing o'er the vale,"—

would carry him and his colony, luggage and live stock, over that long and weary march of fourteen days, in four short hours. As it rushes over its iron track, with wings of fire, he is not in sober Newtown but in a land of Genii, enchanters, magicians! In examining more closely, the snorting, fire-breathing monster, which drags this immense load, at such marvelous speed, he recalls in vain the hydras and dragons, the gorgons and chimeras dire, of his classic reading, for its type or parallel, and with all his senses confused, and his previous knowledge and associations, utterly confounded and bewildered, by its preternatural ferocity and strength, he can only lift his hands in holy horror, as he thinks of the terrible beast in the Apocalypse, or apostrophize it in the language of one of his devout descendants, "Hell in a Harness!"

Instead of the bark canoe, which skimmed its placid waters, when he last gazed on the loveliest of rivers; instead of the tall galleon—spoil of vanquished Spain,—or those huge English war-dogs, that under the great Lord Protector, plucked the laurels of Van Tromp,—some of which he may have seen, ere he left his island home, he beholds sea-palaces, miracles of naval architecture, spreading no canvass to the favoring breeze, but puffing, whistling, screaming, and belching volcano-like smoke and flame, they walk the waters, with incredible velocity and power, moving backwards, forwards, in all directions, turning, tacking, veering, defiant alike of wind, waves, and current, and of all sailing rules, invented by merely mortal mariner. They upset the infallible certainties of science, violate the laws of nature, run counter to all his experience, and gazing upon them, as he would upon Orion or the Pleiades, the infinite awe and reverence of the man can find but this utterance—"How wonderful are thy works, in wisdom Thou hast made them all."

On all sides he would see edifices, that in their vastness, remind him of the monuments of Egyptian or Roman grandeur, but not reared as mausoleums to kings, or votive offerings to the Gods, but temples of industry devoted to the welfare of man, and in their spacious halls, he would see the benignant genius of machinery, "summon around him his willing myriads, assign to each the regulated task, substituting for painful muscular effort, on their part, the energies of his own gigantic arm, and demanding in return, only attention and dexterity, to correct such little aberrations, as occasionally occur in his workmanship."

In the whole circle of modern improvements, there is nothing to which he bends with such profound homage, as the magical powers of machinery. He is dumb with amazement to find, that matter can be so animated by genius, as to perform with skill, exactness, and precision, all the various, complicated, and delicate functions, of the human hand, alike dexterous, in weaving the most gossamer-like threads, in whittling and shaping, to any conceivable form, the hardest metals, in rending rocks, crimping paper, and sticking a pin. With what emotions he would bend over a power-loom, or a sewing-machine, or a screw-machine, or a steam-plough, or wonder of wonders, a "Steam Paddy," grubbing, digging, picking, excavating, embanking, performing most of the functions of its animate name-sake.

When he sees too, that the gross and sensualizing tendencies, of so much material prosperity, are chastened by the cultivation of the ideal, and by the refining influences of the liberal arts, and that Churches, Colleges, Schools, Academies of Painting and Sculpture, follow the pathway of the Steam-Engine and the Power-Loom ; when he sees too, that the guardians both of man's spiritual and material interests, are here working in harmonious coöperation, and that along the iron track of the railroad, and the iron wires of the telegraph, run not only matter and mind, but the eternal plans of Providence, everlasting Truth and Right; when he sees that the poor, tender, persecuted vine which, with so many prayers, and tears, and sad forebodings, he had planted in the wilderness, has been watched, and watered, and guided, by the God of the desolate and oppressed, the stern heart of the old Puritan, melting into joy and gratitude, breaks forth, in the exulting imagery of the Psalmist's pæon,

"Thou brought a vine out of Egypt, Thou cast out the heathen and planted it, Thou madest room for it and it filled the earth. The hills were covered with its shadow, and the boughs thereof, were like the goodly cedar trees. She stretched forth her branches unto the sea, and her boughs to the river."

The world is not yet finished! God has indeed completed his six days' work, but the dawn is now purpling the East, in the first, of man's countless days of labor.

The Future opens before us with the most auspicious omens, and in inconceivable grandeur. We can not indeed lift the veil, but the analogies of the past, throw such rays and scintillations into the void, that we can for a moment fathom its depths and penetrate its gloom. All inventions and discoveries are procreative. The spinning wonder of Arkwright speedily begat the weaving wonder of Cartwright.

The railroad is the legitimate progeny of the steamboat. The Ambrotype, the Photograph and sun-painting in colors, were all within the grasp of time and patience, when Daguerre succeeded in arresting, the fleeting image of the Camera. Every year now, adds more to the usefulness of steam, than the two hundred years, that separated Watt from Worcester. It is thus that every new conquest becomes a point of vantage gained—a parallel carried—a Malakoff taken, from whence new batteries can be directed, and redoubled energies turned, upon the undiscovered and unknown. All the science too of the world, and a large proportion of its intellect, at this era, culminate upon the single point of practical aggrandizement. Astronomy is not now calculating nativities, while the mariner is struggling in the breakers, nor chemistry using quicksilver as an amalgum only, while the miner and metallurgist are separating the ore, by wearisome processes, nor mechanical philosophy flying off in search of the perpetual motion, while the cotton-planter picks out the seed from the fiber, with his fingers.

In the light of such facts and analogies, and in that of the progressive law of coöperating Thought and Labor, and of past experience, we repose securely upon the conviction, that the realities of THE FUTURE, will immeasureably transcend the wildest dreams of prophecy. We seem sailing along the very borders, hearing the ripple that breaks upon the shore, where the USEFUL ARTS are finally to be installed, in absolute and immutable perfection. The cloud lifts, and we catch, for a moment, a glance at the wonders beyond. We see meadows and pastures green, oppressed with unknown varieties of vegetable life, the barren places of the earth, rejoicing in such crops, as the Nile might have witnessed, in those seven years of plenty. The air is navigated, the mountains leveled, great arms of the sea bridged with cables or spanned by arches.

We strive in vain, to estimate the infinite progeny, that have sprung from some well-known invention. We see heaped together, in all the beautiful and fantastic shapes of a magical Architecture, piles upon piles, of granite, marble, and iron, towering upward to the clouds, in their vastness, resembling the mountainous upheavals of the convulsed earth, and in splendor and gorgeousness, that celestial city which Bunyan saw in vision. We stand beneath domes, compared with which, St. Peter's is as an emmet's tumulus, and before palaces, in whose vestibule the Vatican might be lost. We guage and measure the capacities, of new powers and forces added to the resources of man, beside which Steam is weak and Electricity slow. We behold machinery so marvelous in strength, adaptation, and economy, that the steam-engine has been driven into the museum of the antiquary, and the spinning-jenny discarded, as a toy of childhood. We see curious workmanship wrought from new metals, or new combinations of the old ones, and beautiful fabrics, fitting in and supplying wants and necessities, now unanswered, and all woven from natural products, as much unknown to us, as cotton to the Saxons, or caoutchouc to the Chaldeans. Winding around an immense plaza, sparkling with jets and fountains, here somber with the umbrage of unknown monarchs of the wood, there gay, gaudy and fragrant with a tropical vegetation, surpassing an Epicurean's dream, and everywhere adorned with the colossal images of immortal inventors and contrivers, we see, (celebrating some high festival of industry, or inaugurating an Agricultural Fair, in the 25th century,) a triumphal procession, jubilant with the shoutings and rejoicings of emancipated labor, radiant with ruby and amethyst, with diamonds and pearls, the spoils of the sea and the land, with the crowns, trophies, and ensigns of vanquished nature, with the emblems of new arts, manufactures, and handicrafts, and proudly preëminent among them all, though chained to the victorious car, the terrific elements that have been discovered and enslaved; and on the architrave of the vast temple that flings open its golden gates, to receive the exulting multitude we read, as an incentive to industry and enterprise even in the presence of seeming perfection—

"WORK ON, WORK EVER—THE WORLD IS NOT YET FINISHED."

www.ingramcontent.com/pod-product-compliance
Lightning Source LLC
LaVergne TN
LVHW020741120826
845150LV00010B/2285